AF877842

The Adventures of Bella Noelle

Hey guys it's Bella and I am so glad you decided to join me on another adventure. In this workbook, we will learn about shapes. We will trace our shapes, color our shapes, write the names of our shapes and so much more. So come on guys let's get going on our adventure!

Coloring Shapes

Circle

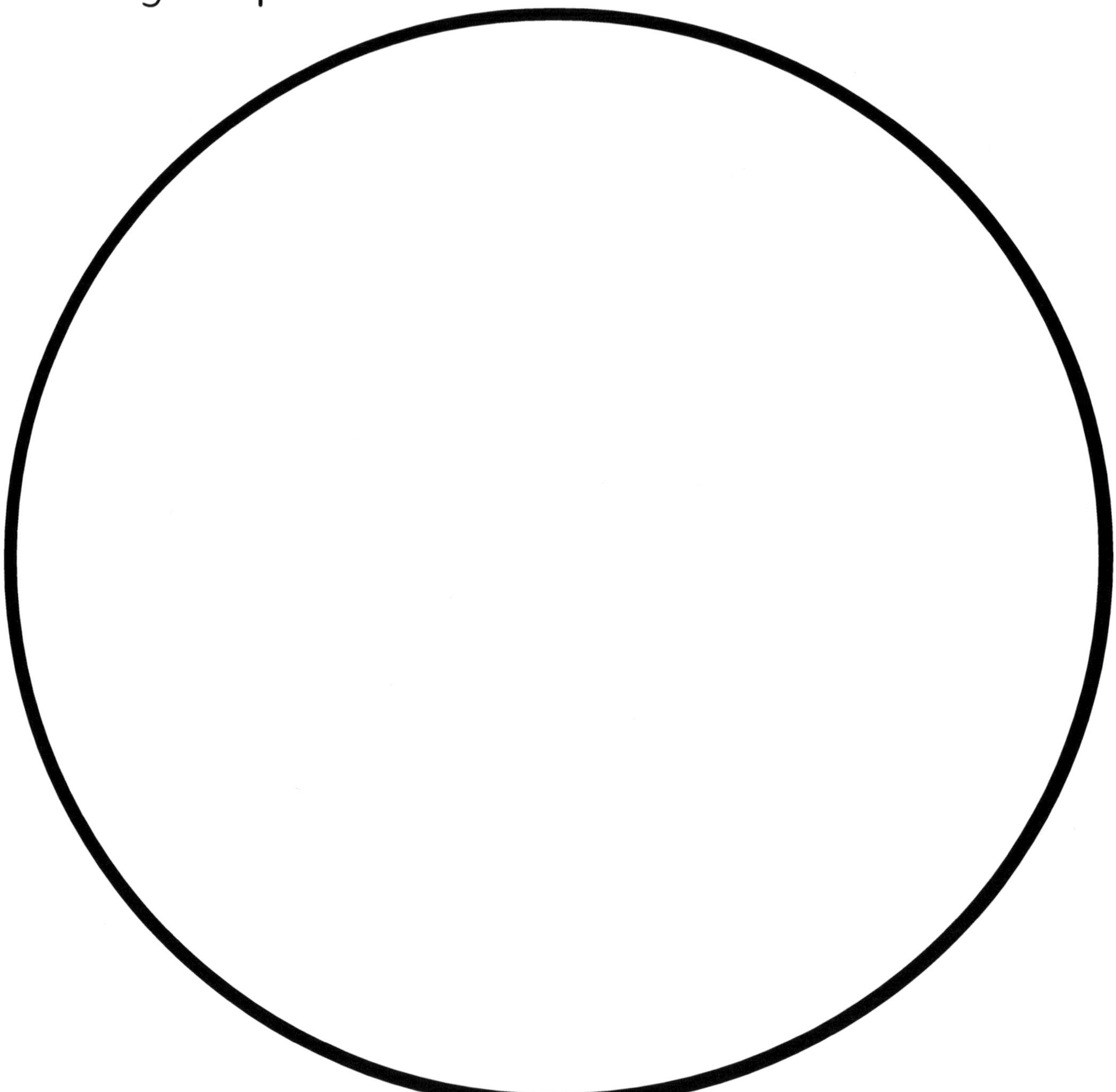

Tracing Shapes

Circle

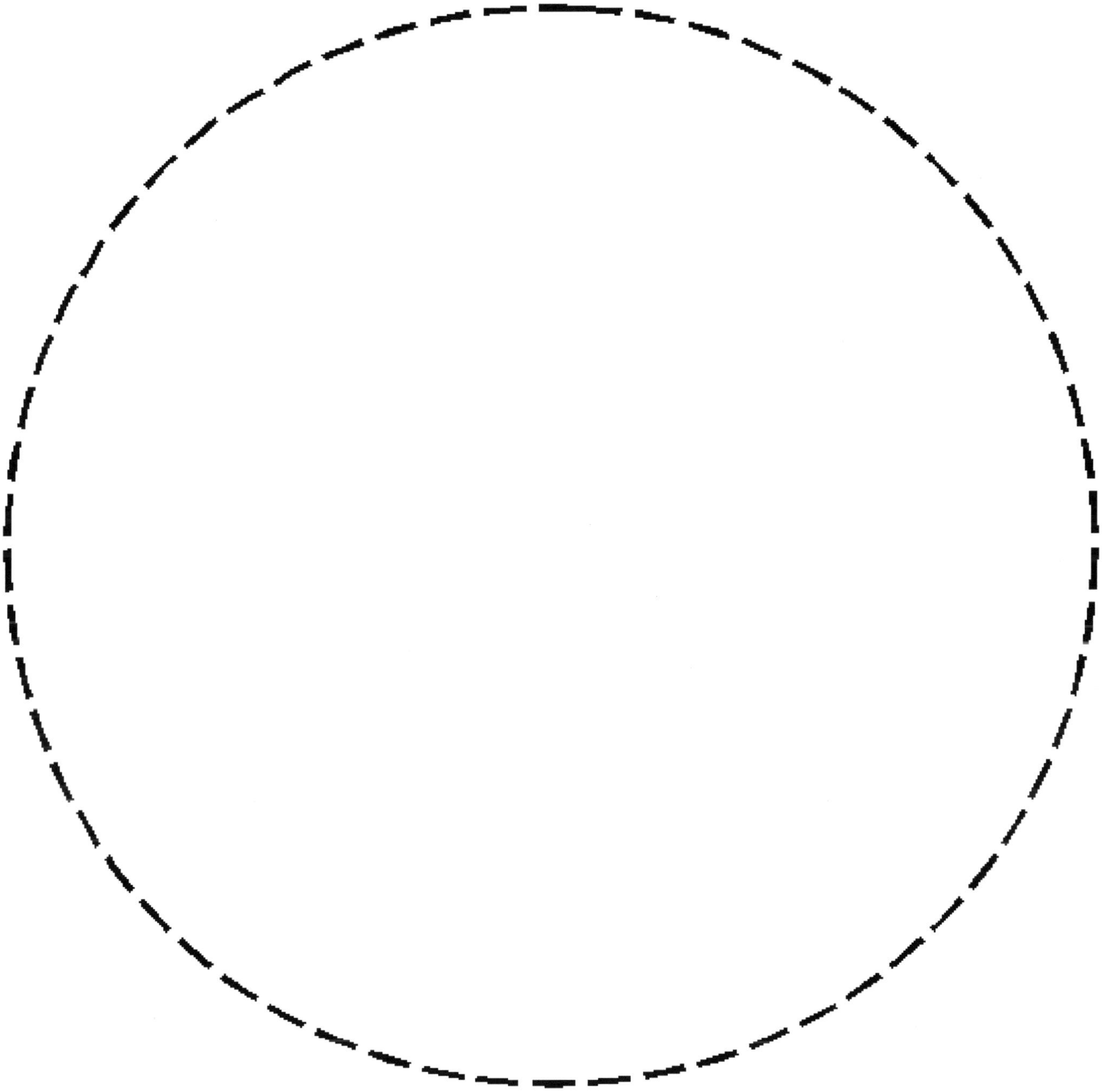

Circle

Circle

Writing Shape Words

Circle

circle circle circle

circle circle circle

circle circle circle

circle circle circle

circle circle circle

circle circle circle

Circle

Name

CIRCLE

Circle: a figure bounded by one curved line, and such that all straight lines drawn from a certain point within it to the bounding line, are equal.

Directions: Color the circles red in the shapes shown below.

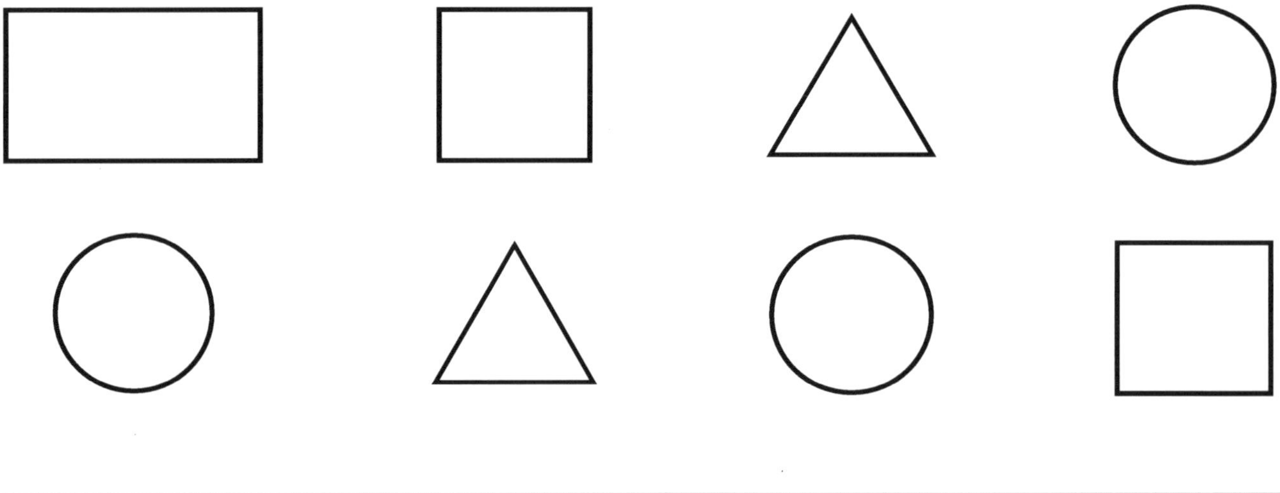

Practice writing the word circle and then draw a circle below (a shape that has one continuous curved line) and then decorate it using your crayons.

circle

Patterns

Circle the shape that comes next.

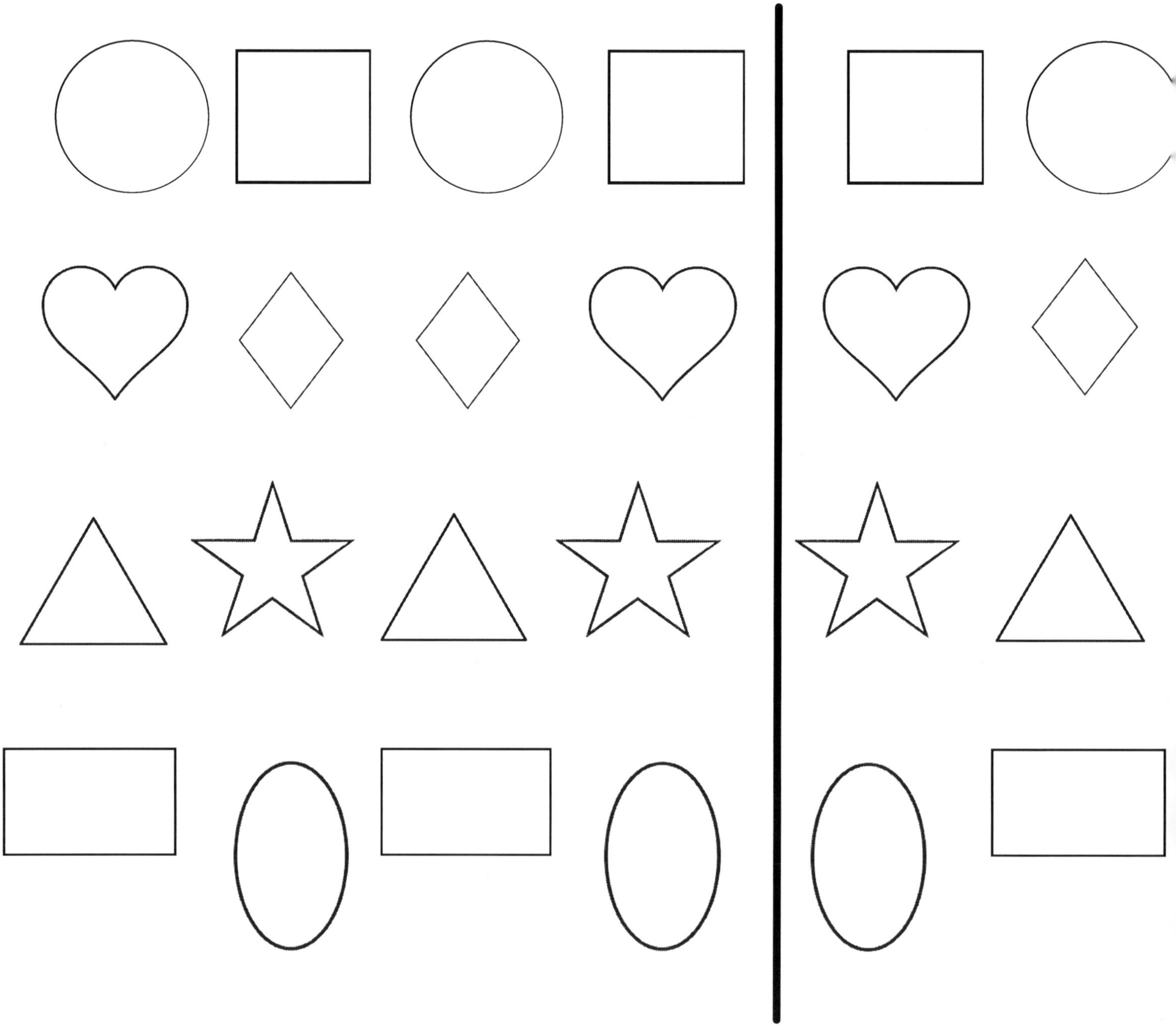

Coloring Shapes Square

Tracing Shapes

Square

Square

Square

Writing Shape Words

Square

square square square

square square square

square square square

square square square

square square square

square square square

square square square

Square

Coloring Shapes

Name

SQUARE

Square: a shape with four equal straight sides and four right angles

Directions: Color the squares blue in the shapes shown below.

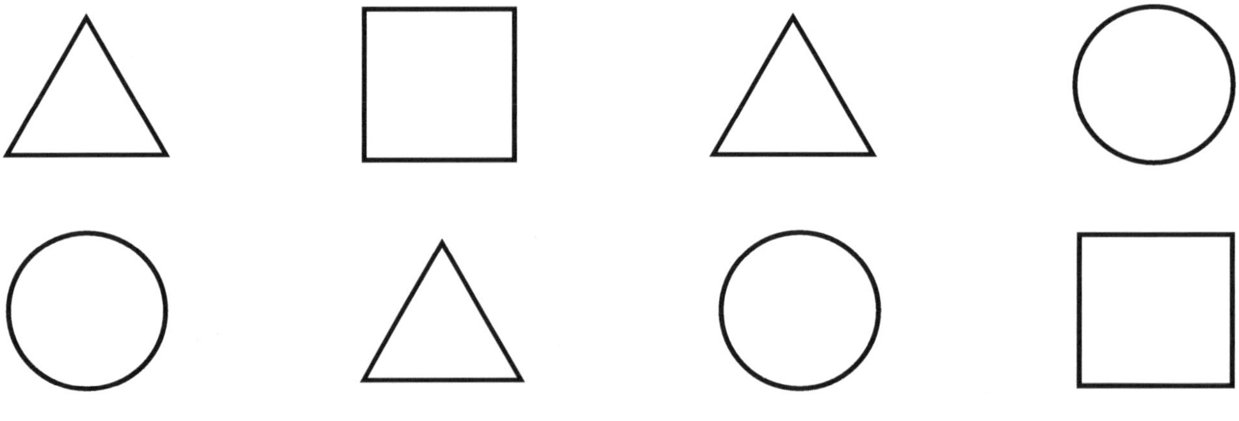

Practice writing the word square and then draw a square below (a shape that has four equal straight sides and four right angles) and then decorate it using your crayons.

square

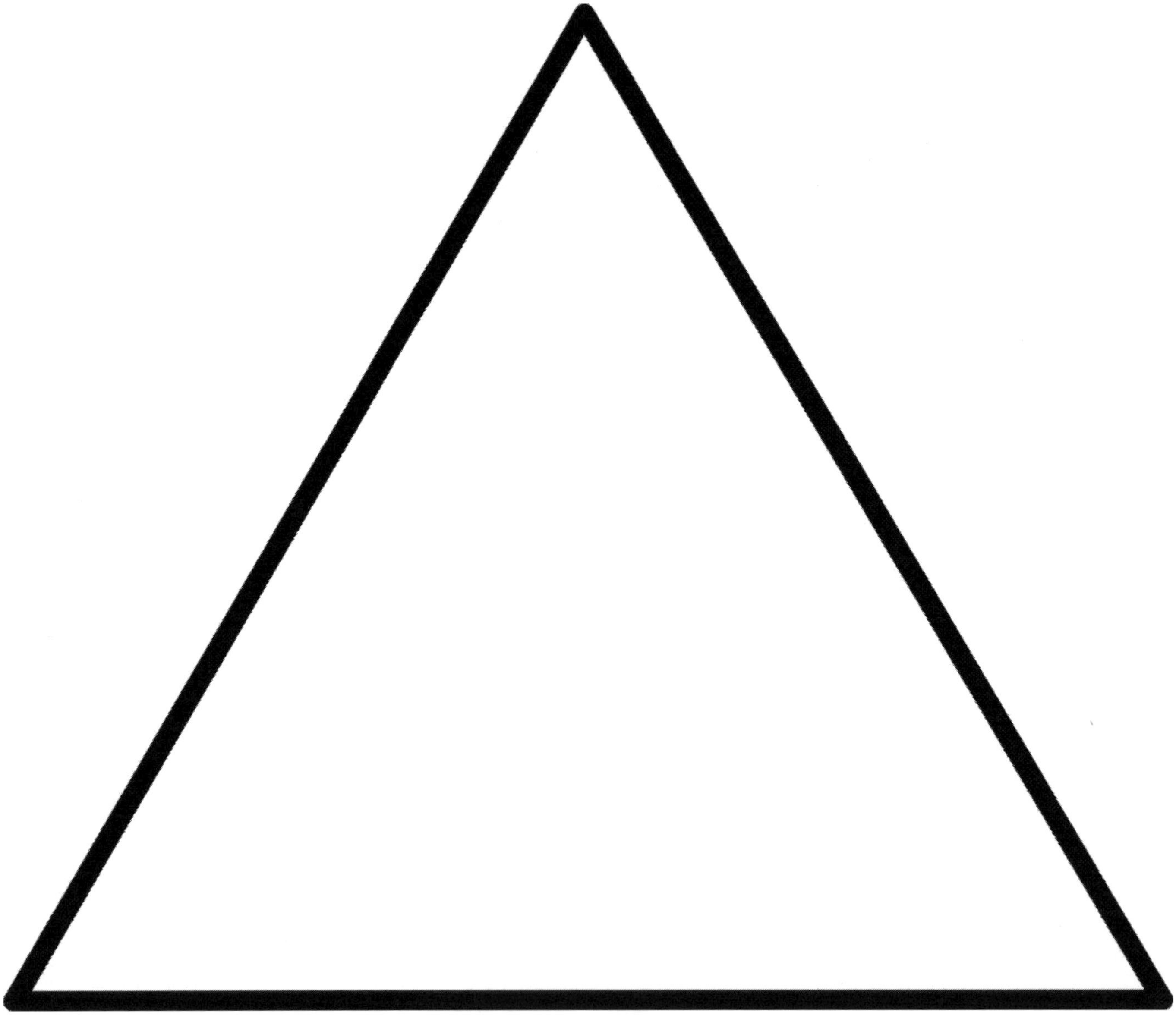

Tracing Shapes

Triangle

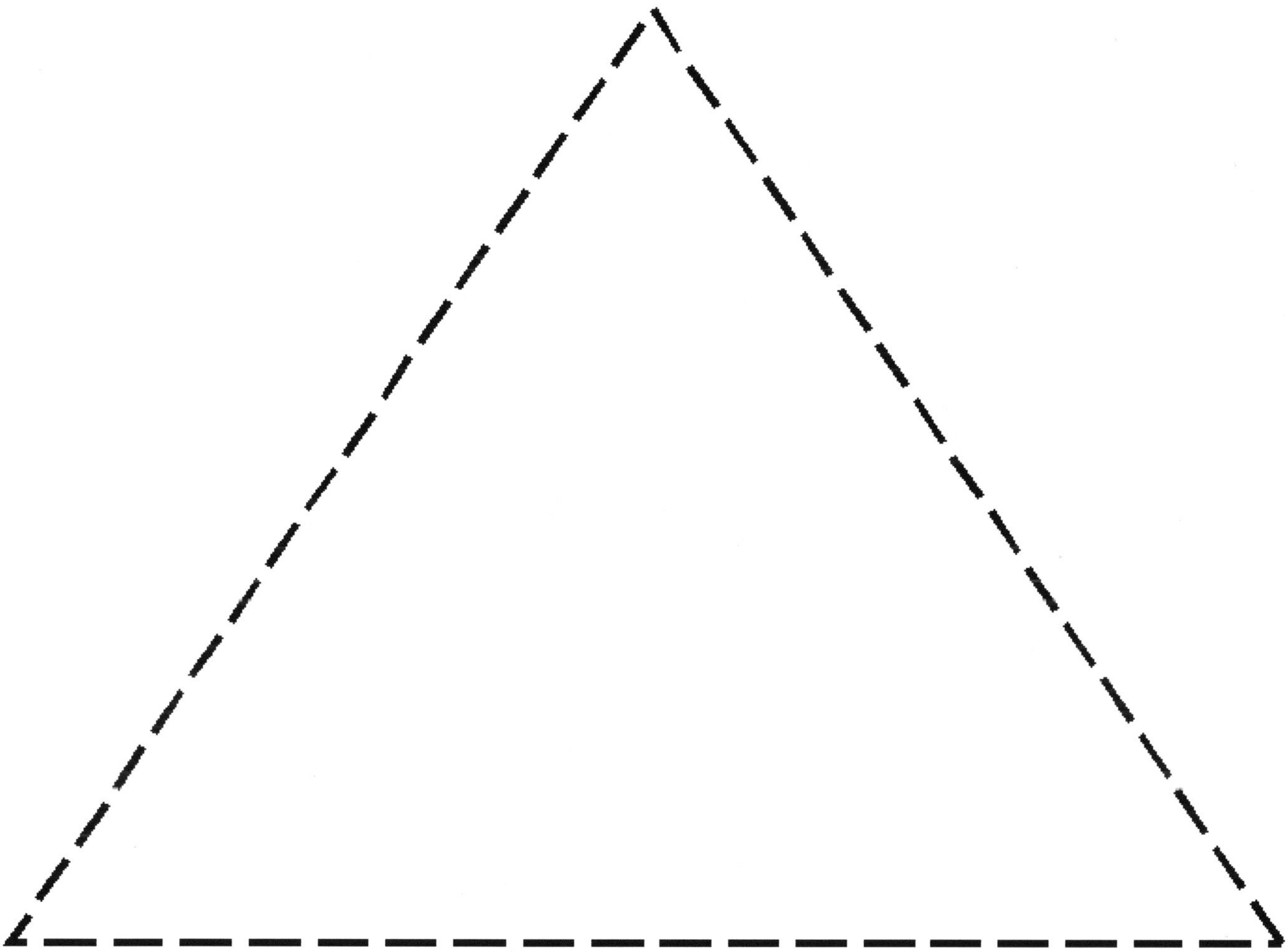

Triangle

Triangle

Writing Shape Words

Triangle

triangle triangle triangle

triangle triangle triangle

triangle triangle triangle

triangle triangle triangle

triangle triangle triangle

triangle triangle triangle

triangle triangle triangle

triangle triangle triangle

Name

TRIANGLE

Triangle: a shape with three straight sides and three angles.

Directions: Color the triangles green in the shapes shown below.

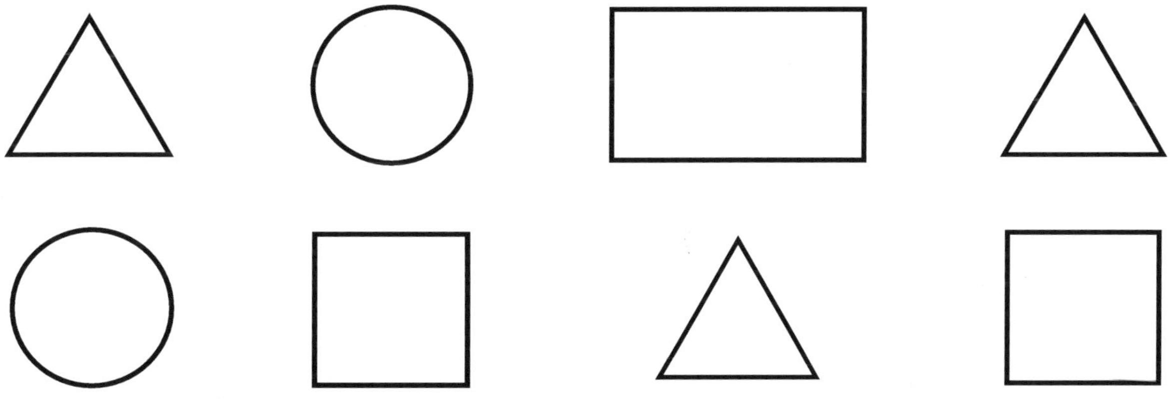

Practice writing the word triangle and then draw a triangle below (a shape with three straight sides and three angles) and then decorate it using your crayons.

triangle

Coloring Shapes

Rectangle

Tracing Shapes

Rectangle

Rectangle

Rectangle

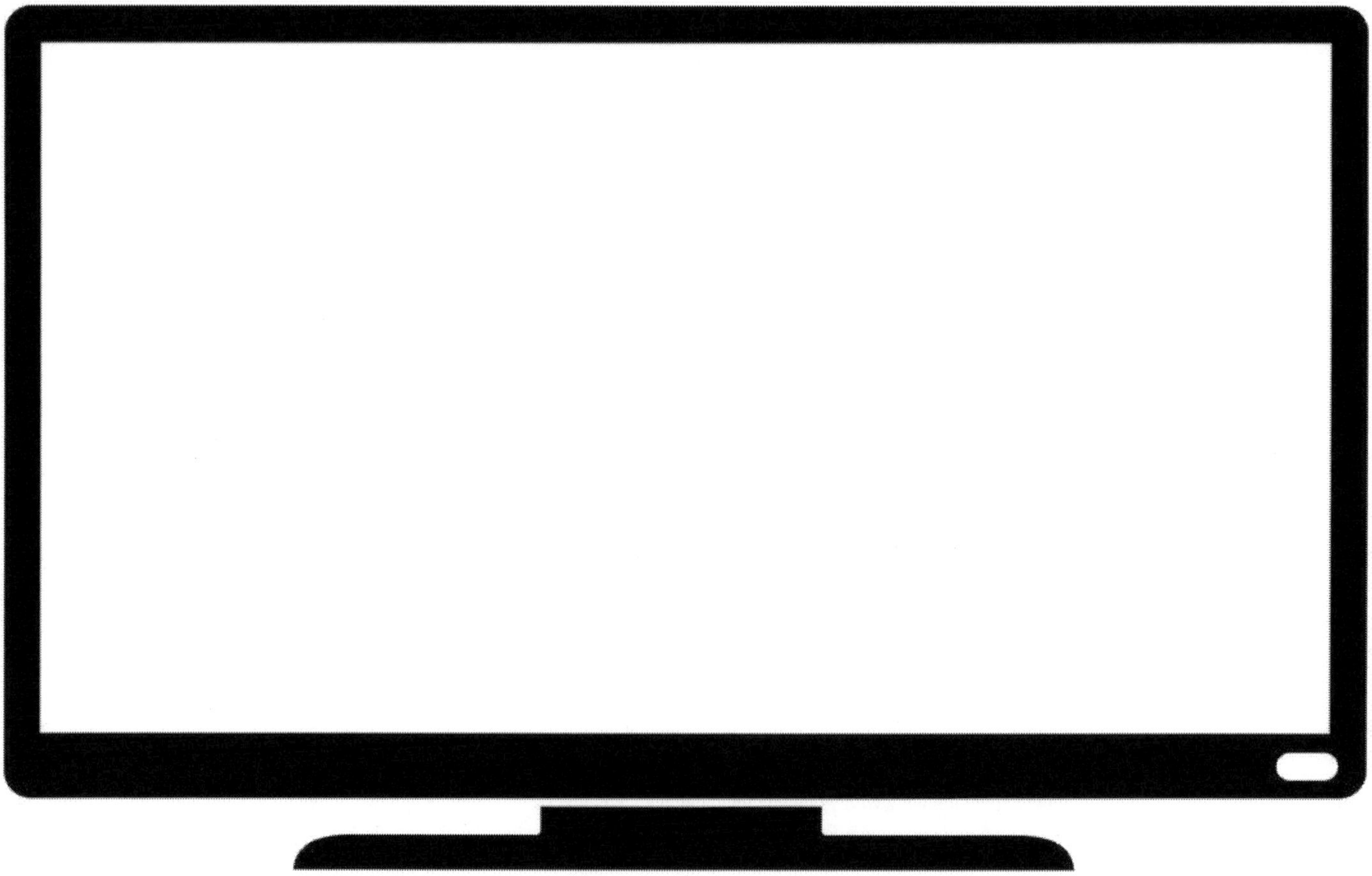

Rhombus

Tracing Shapes

Rhombus

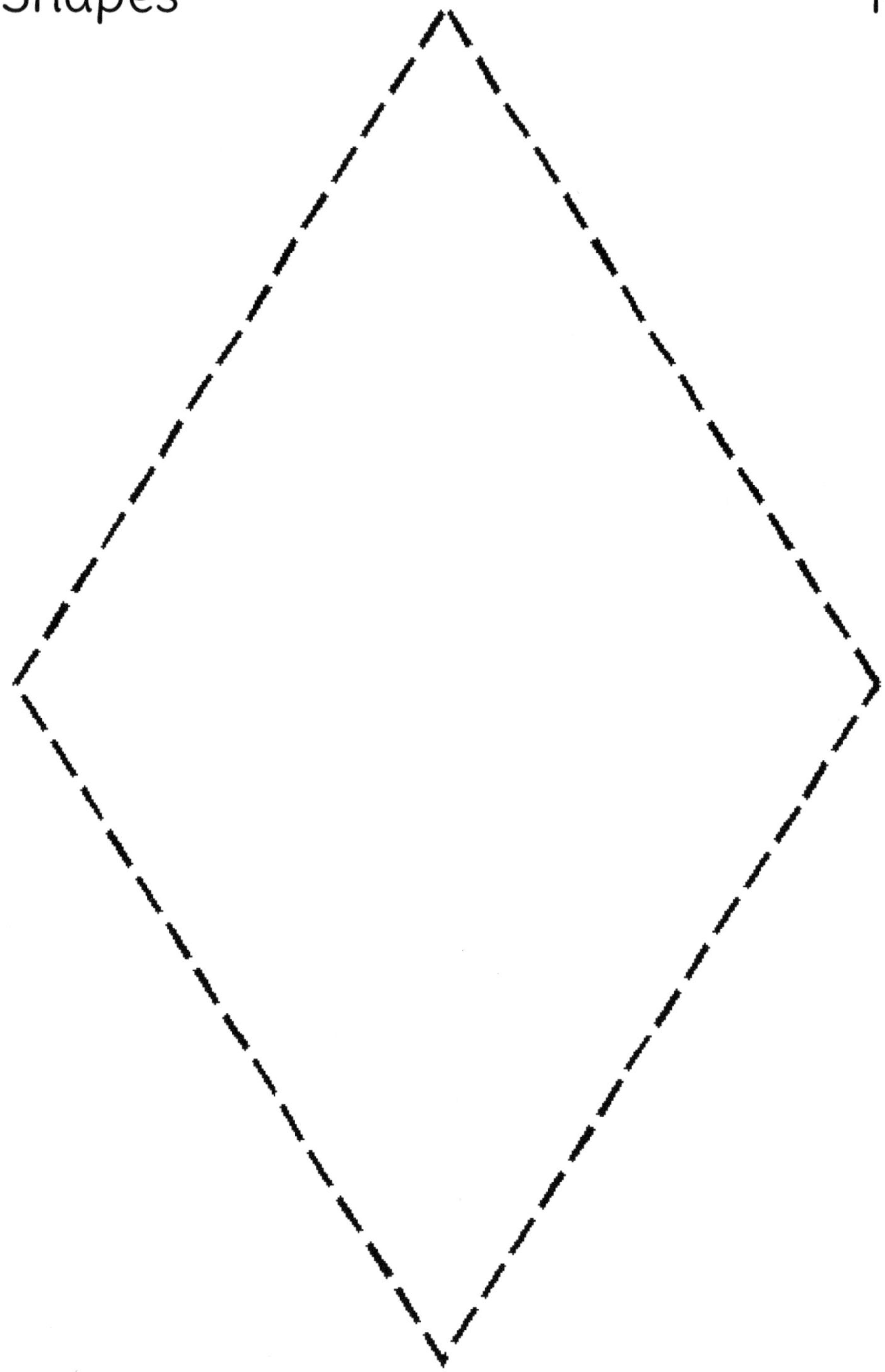

Coloring Shapes

Rhombus

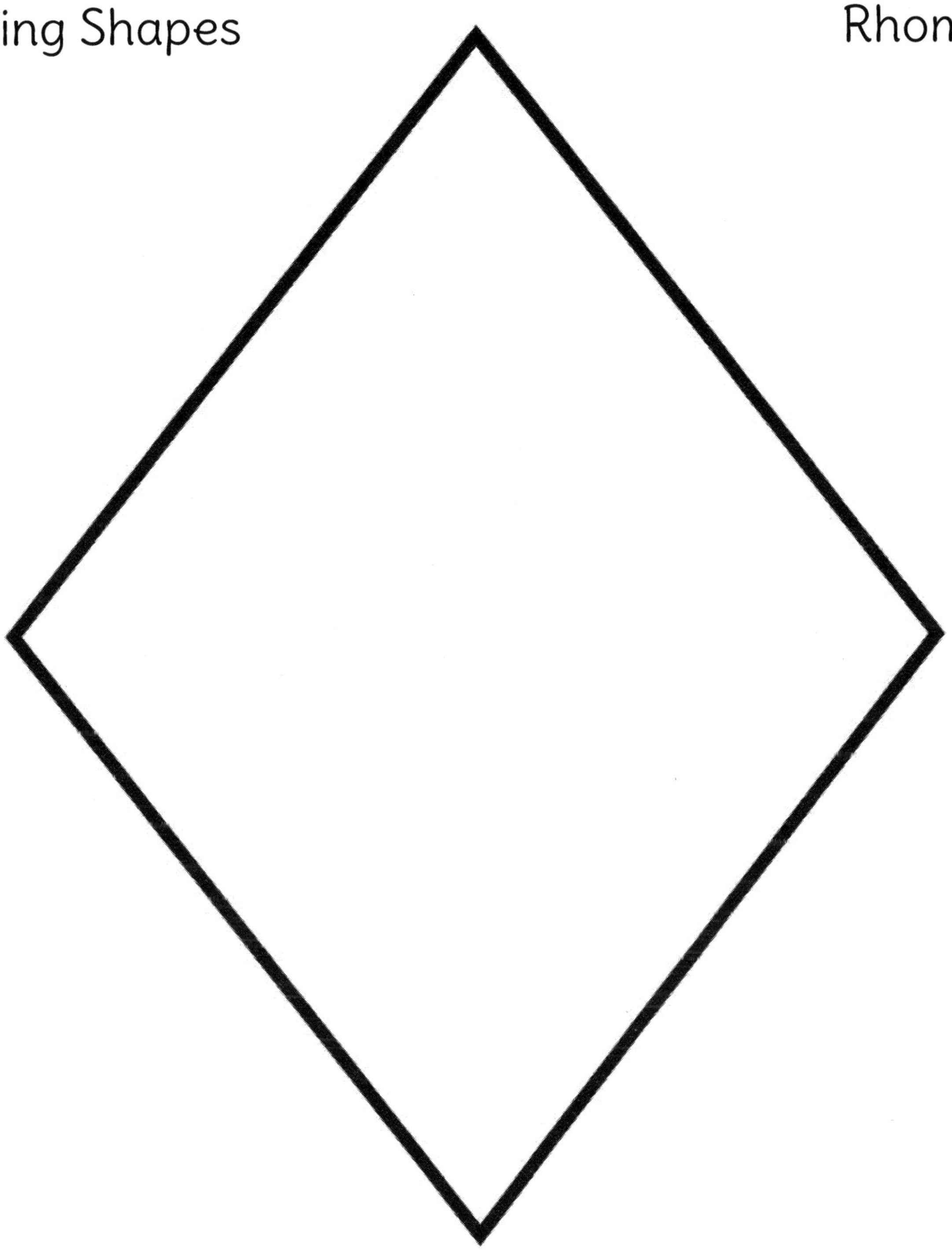

Practice writing the word rectangle and then draw a rectangle below (a shape with four straight sides and four right angles, with unequal adjacent sides) and then decorate it using your crayons.

rectangle

Name

RECTANGLE

Rectangle: a shape with four straight sides and four right angles, with unequal adjacent sides

Directions: Color the rectangles orange in the shapes shown below.

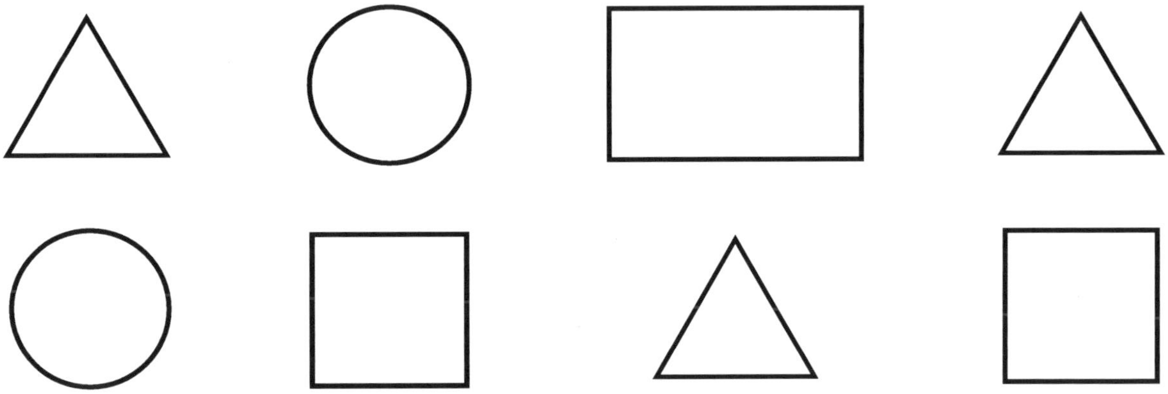

Writing Shape Words

Rectangle

rectangle rectangle

rectangle rectangle

rectangle rectangle

rectangle rectangle

rectangle rectangle

rectangle rectangle

Writing Shape Words

Rhombus

rhombus rhombus

rhombus rhombus

rhombus rhombus

rhombus rhombus

rhombus rhombus

rhombus rhombus

Name

DIAMOND

Diamond: a shape with four straight sides of equal length forming two opposite acute angles and two opposite obtuse angles

Directions: Color the diamonds purple in the shapes shown below.

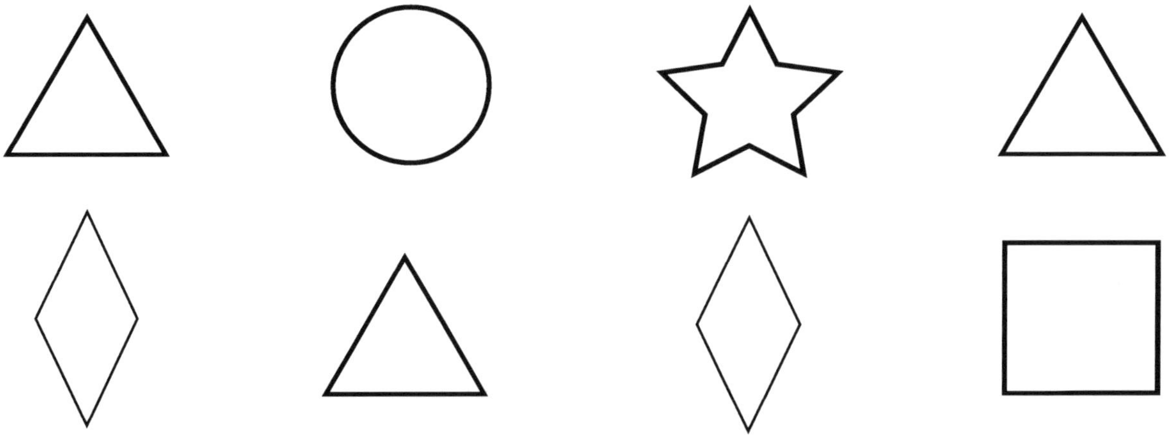

Practice writing the word diamond and then draw a diamond below (a shape with four straight sides of equal length forming two opposite acute angles and two opposite obtuse angles) and then decorate it using your crayons.

diamond

Rhombus

Coloring Shapes Star

Tracing Shapes

Star

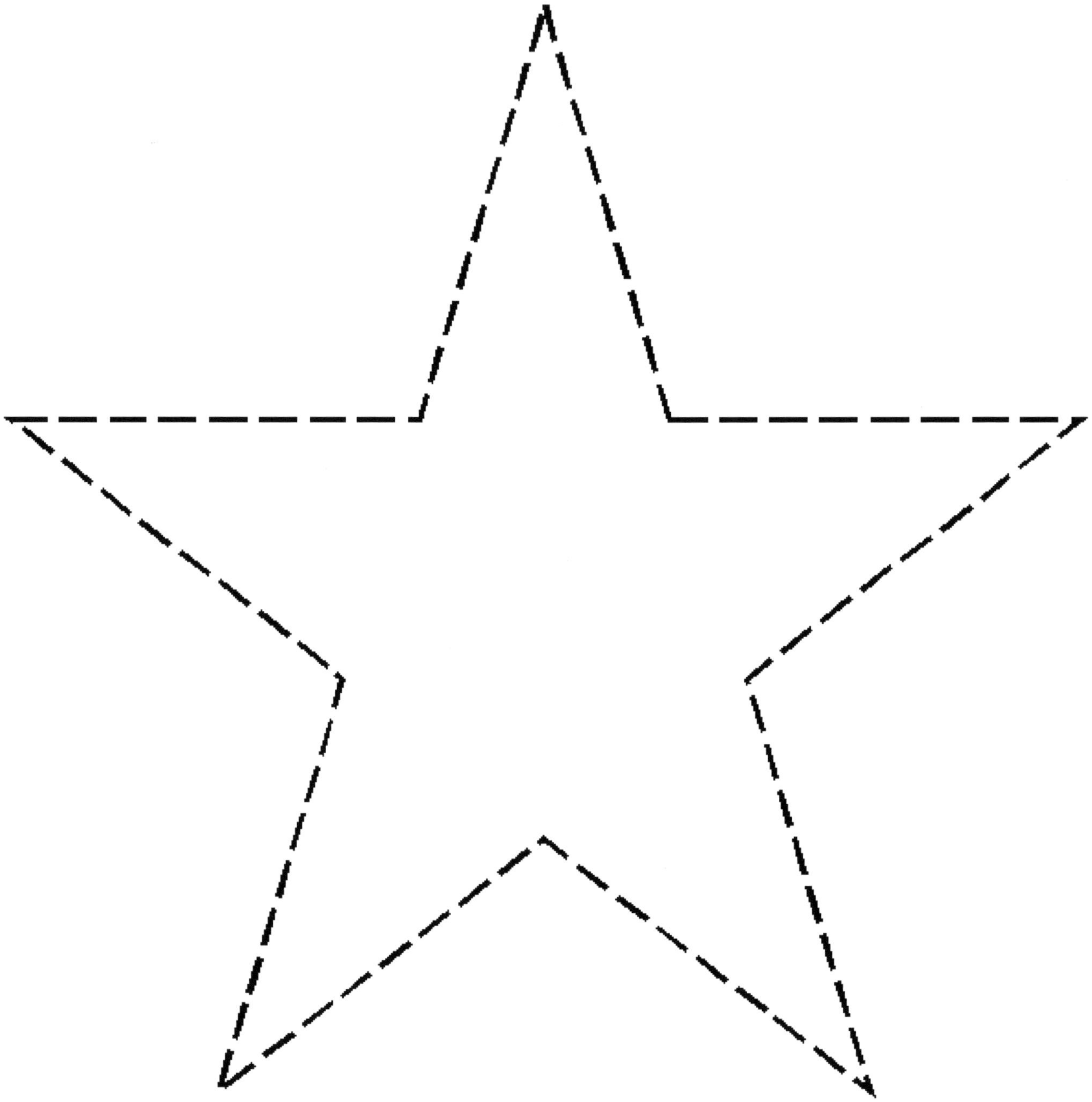

Star

Star

star star star star

star star star star

star star star star

star star star star

star star star star

star star star star

Star

Name

STAR

Star: a shape with four or more points that looks like a star that might be found in the night sky

Directions: Color the stars blue in the shapes shown below.

Practice writing the word star and then draw a star below (a shape with four or more points that looks like a star that might be found in the night sky) and then decorate it using your crayons.

star star

Word Search

Find the shape words from the word bank.

Z	R	E	C	T	A	N	G	L	E
R	L	O	G	S	U	V	E	T	T
F	A	J	I	P	Q	L	M	R	R
R	I	T	L	M	C	U	I	R	U
U	M	X	S	R	D	A	A	D	E
U	X	L	I	T	N	B	W	R	Q
O	S	C	R	G	O	V	A	L	E
Y	G	A	L	G	W	S	P	X	M
Q	E	E	S	U	B	M	O	H	R
H	C	W	L	O	Y	G	C	D	V

Word Bank

Circle	Triangle	Rhombus	Oval
Square	Rectangle	Star	Heart

Coloring Shapes

Heart

Tracing Shapes Heart

Coloring Hearts

Heart

Heart

Writing Shape Words

Heart

heart heart heart

heart heart heart

heart heart heart

heart heart heart

heart heart heart

heart heart heart

Name

HEART

Heart: a conventional representation of a heart with two equal curves meeting at a point at the bottom

Directions: Color the hearts red in the shapes shown below.

Practice writing the word heart and then draw a heart below (a conventional representation of a heart with two equal curves meeting at a point at the bottom) and then decorate it using your crayons.

heart

Coloring Shapes

Oval

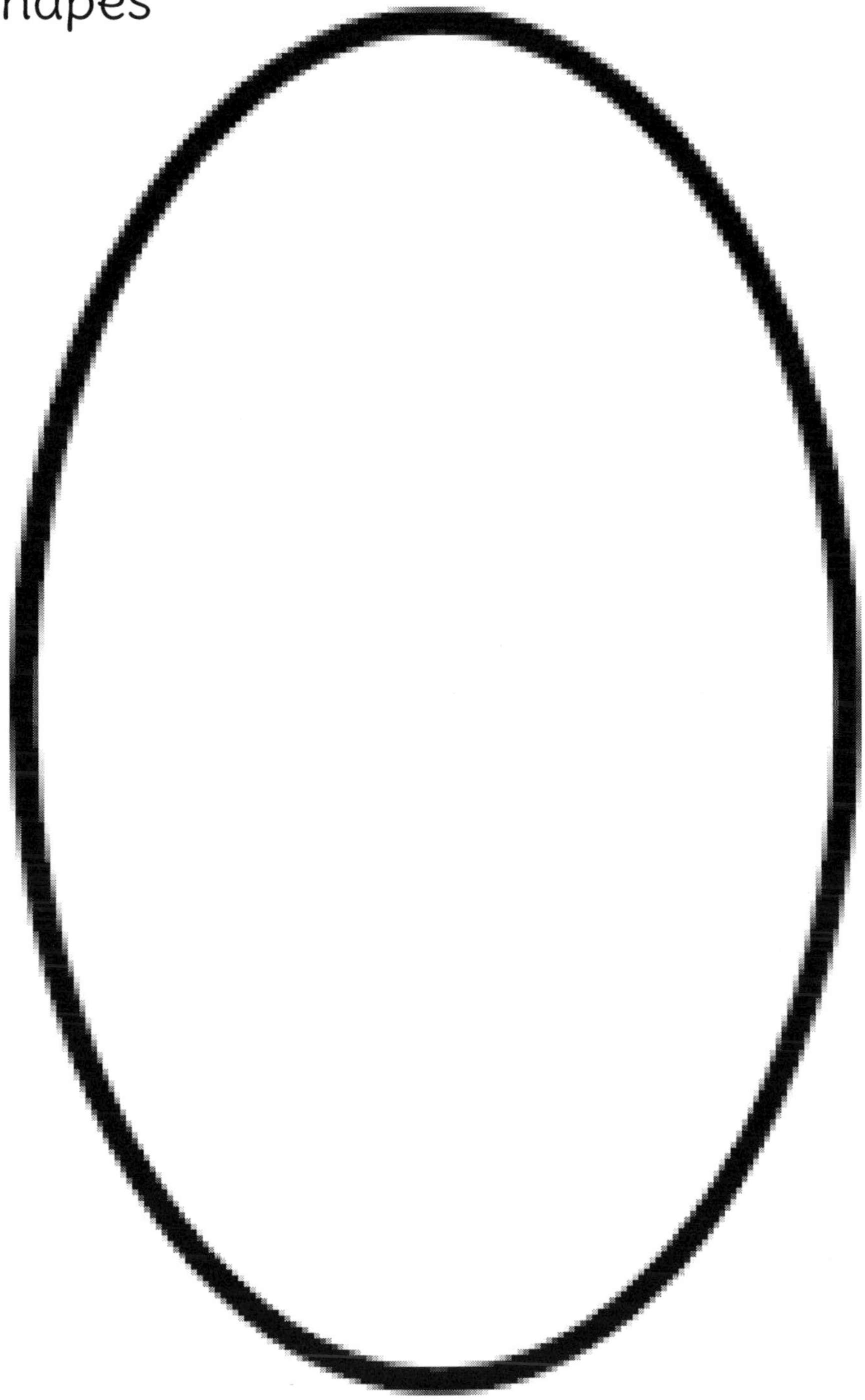

Tracing Shapes

Oval

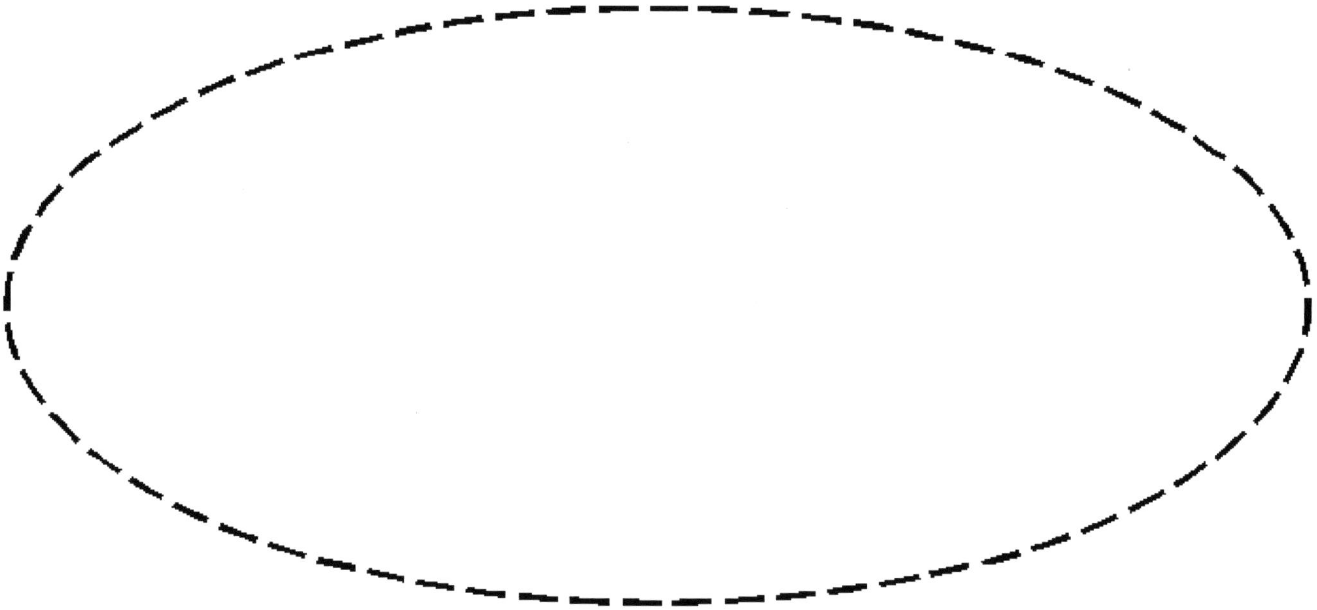

Writing Shape Words Oval

oval oval oval oval

oval oval oval oval

oval oval oval oval

oval oval oval oval

oval oval oval oval

oval oval oval oval

Oval

Oval

Coloring Shapes

Name

OVAL

Oval: a shape having a rounded and slightly elongated outline or shape like that of an egg

Directions: Color the ovals pink in the shapes shown below.

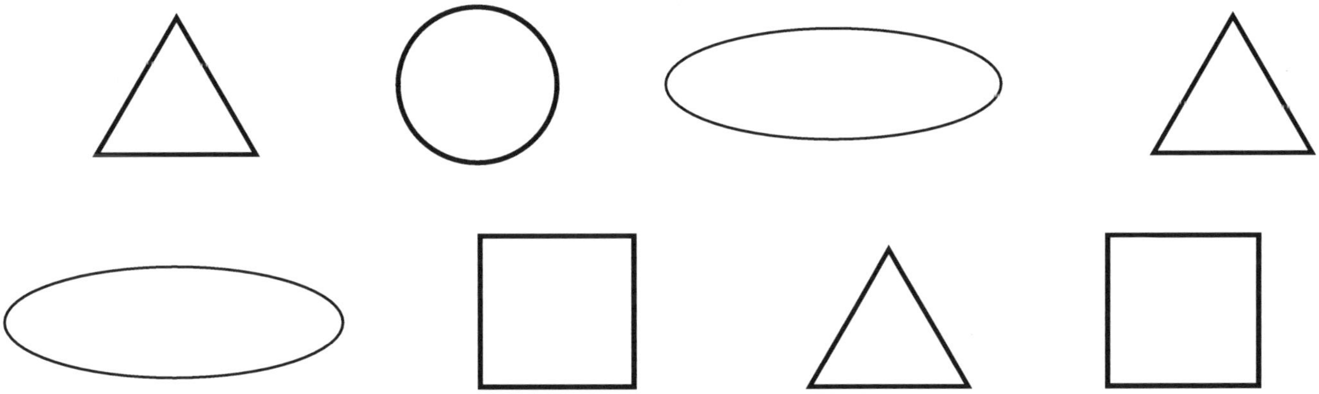

Practice writing the word oval and then draw an oval below (a shape having a rounded and slightly elongated outline or shape like that of an egg) and then decorate it using your crayons.

oval oval

Coloring Shapes

Coloring Shapes

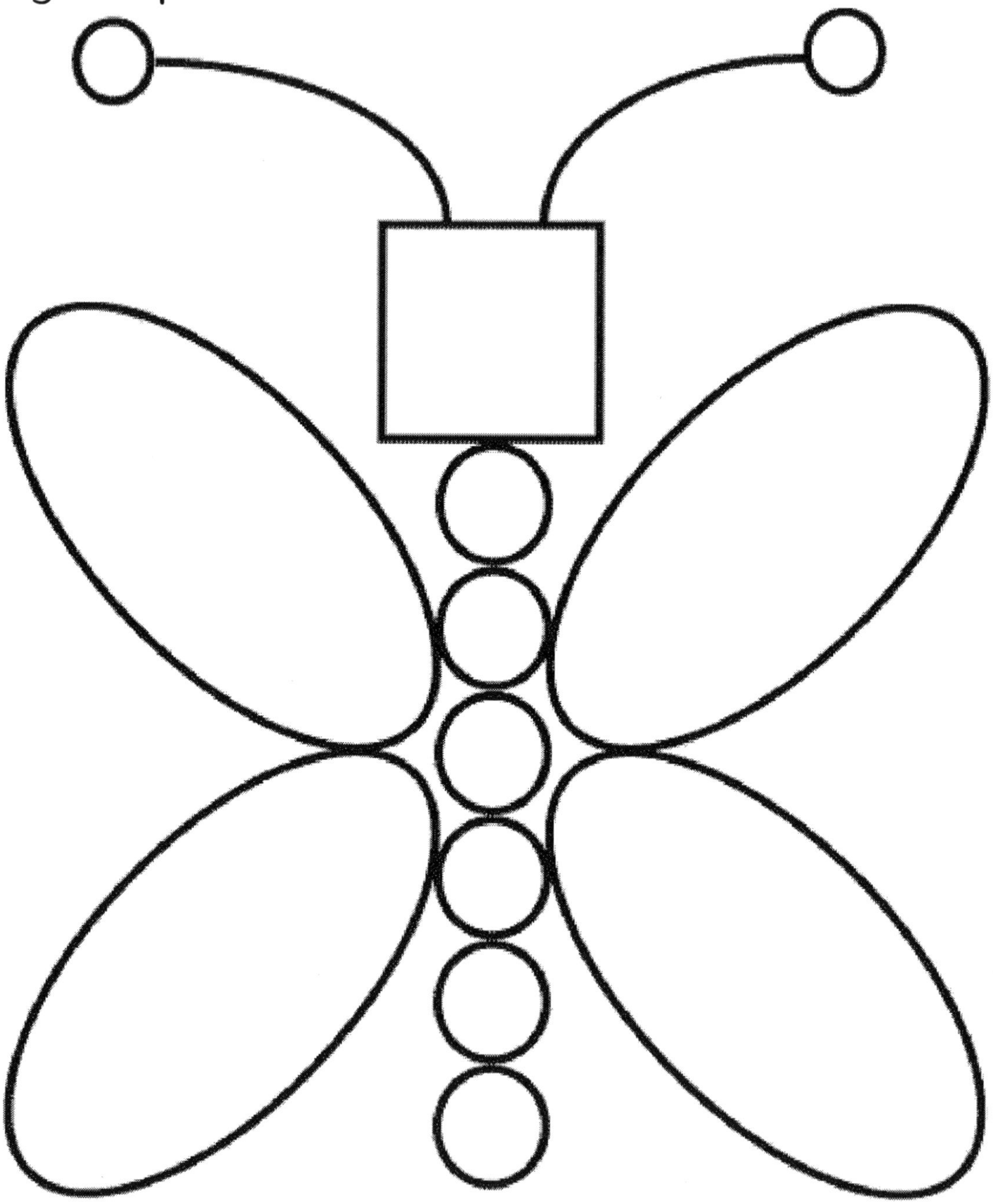

Made in the USA
Coppell, TX
04 November 2026

11282124R00037